THIS BOOK BELONGS TO:

_____

The Wonderful World of Crabs

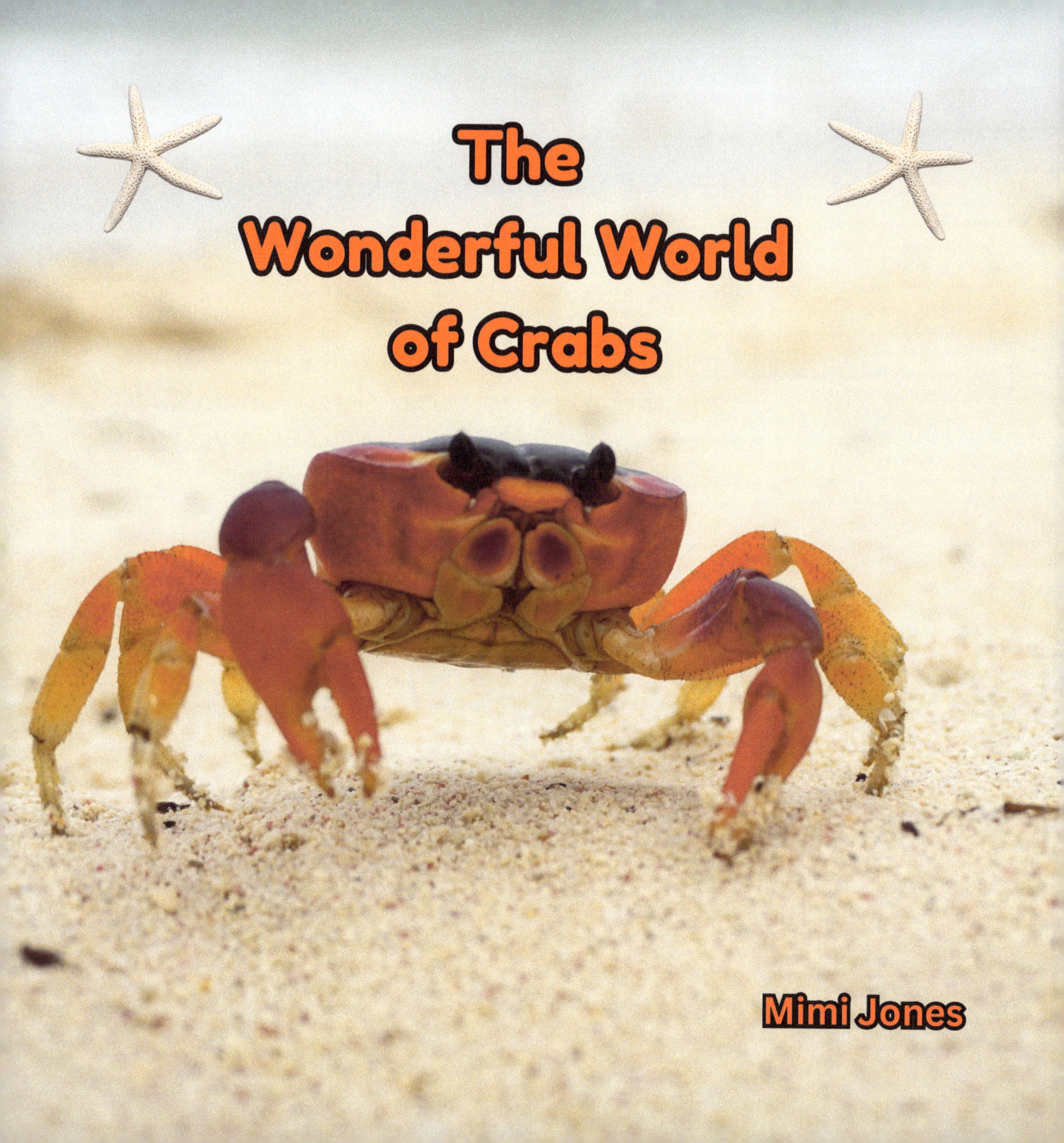

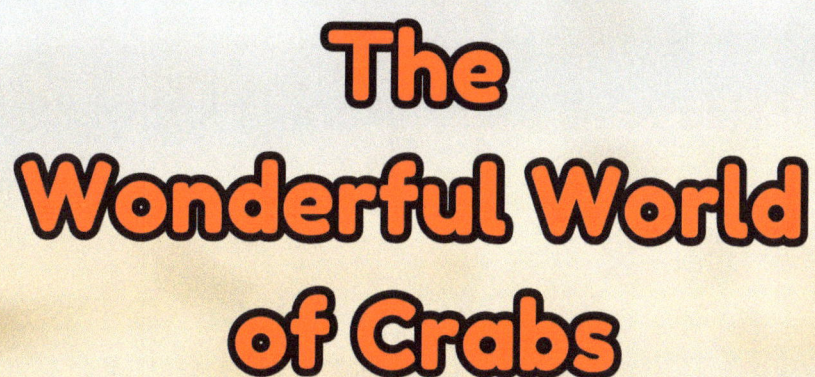

Mimi Jones

Dedicated to all the seekers of knowledge.

All rights reserved.
No part of this book may be reproduced in any form or by any means, electronic or mechanical, and no photocopying or recording, unless you have written permission from the author.

ISBN 978-1-958985-40-3

Text copyright © 2025 by Mimi Jones

www.joeysavestheday.com

A Mimi Book

10

Crabs belong to the order Decapoda, which means "ten-footed."

The Sally Lightfoot crab is found primarily in the Galapagos Islands.

Crabs can be found in all of the world's oceans, as well as on land and in freshwater.

1. Pacific Ocean
2. Atlantic Ocean
3. Indian Ocean
4. Southern Ocean
5. Arctic Ocean

The largest species of crab is the Japanese spider crab, with a leg span of up to 12 feet.

12

The smallest crab species is the pea crab, which measures less than an inch across.

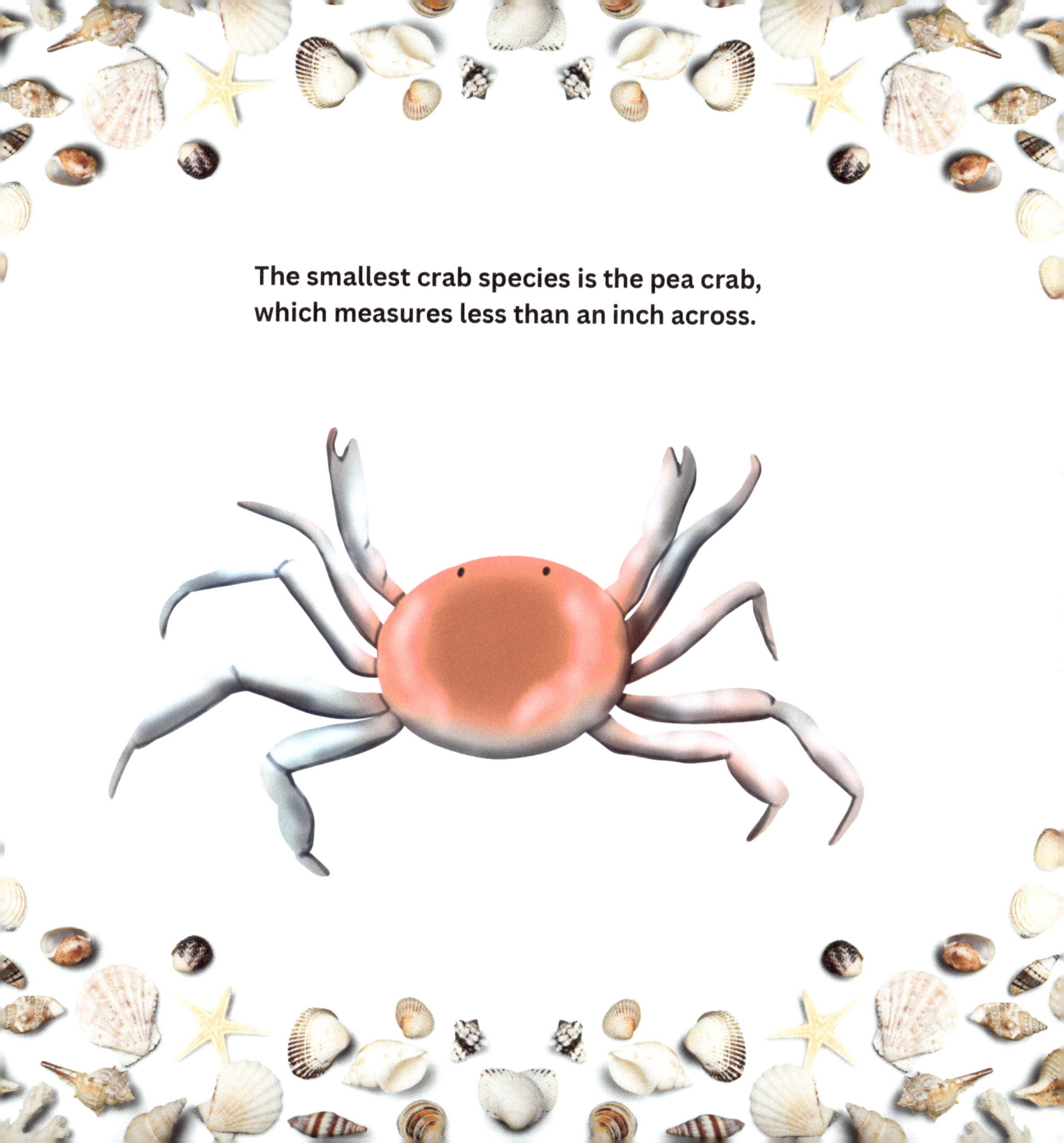

Crabs have a pair of claws called chelae that they use for defense and feeding.

Some crabs, like the fiddler crab, have one claw significantly larger than the other.

Crabs walk sideways because their legs are jointed at an angle that allows for lateral movement.

Crabs have compound eyes on stalks, giving them a wide field of vision.

They can regenerate lost limbs over time.

Crabs communicate with each other using a combination of visual signals and sounds.

Hermit crabs are not true crabs; they belong to a different family called Paguroidea.

Some species of crabs, like the coconut crab, can climb trees.

Some crabs have a specialized appendage called a "swimmeret" to aid in swimming.

Crab swimmerets, located on the back legs, are paddle-shaped appendages that enable the crab to swim efficiently.

Many crabs are omnivores, eating both plants and animals.

Crabs have a unique way of breathing called branchial respiration, using gills.

The horseshoe crab is not a true crab; it belongs to a group of arthropods called chelicerates.

Crabs play a crucial role in their ecosystems, often as scavengers that help clean up the environment.

Some species, like the blue crab, are highly valued as seafood.

Crabs can change color to blend in with their surroundings, a behavior known as camouflage.

SPEED LIMIT 10

The ghost crab can move at speeds of up to 10 miles per hour.

Crabs use their legs to dig burrows in the sand or mud for shelter.

During mating season, some crabs perform complex courtship dances.

Crabs are ancient creatures, with fossils dating back to the Jurassic period, around 200 million years ago.

Count the crabs.

www.ingramcontent.com/pod-product-compliance
Lightning Source LLC
Chambersburg PA
CBHW040029050426
42453CB00002B/59